The cover image is a red panda (aka: lesser panda, red bear-cat, and red cat-bear), a small mammal native to the eastern Himalayas and southwestern China. Classified as endangered, its wild population is estimated at less than 10,000 mature individuals. The red panda, slightly larger than a domestic cat, feeds mainly on bamboo, but also eats eggs, birds, insects, and small mammals. It is a solitary animal, mainly active from dusk to dawn. While the name might suggest otherwise, the red panda is not closely related to the better known, black and white, giant panda. Instead, this cute animal is related to the raccoon, weasel and skunk.

Written by

Brent A. Ford

&

Lucy McCullough Hazlehurst

© 2017 by nVizn Ideas LLC

www.nviznideas.com

The Authors

Brent A. Ford

With more than 30 years experience, Brent is a science educator, education researcher, writer, curriculum developer, professional development leader, and entrepreneur & business owner. Brent is the founder & CEO of nVizn Ideas, LLC.

Lucy McCullough Hazlehurst

With more than 25 years experience, Lucy is a language arts & primary grade educator, writer, curriculum developer, researcher, and leader in all things children's theatre. Lucy is the Creative Director & Chief Writer for nVizn Ideas, LLC.

Instructions for Parents & Teachers

The purpose of this book is to help children develop an appreciation for the world in which we live. We intend to accomplish this goal by providing a series of spectacular images coupled with a variety of verse...from time-honored and classic poetry to new and sometimes humorous verse.

This is not the type of book typically read in one session. We encourage you to use the pictorial Table of Contents to come and go as children ask questions about the world. Children can certainly experience the book on their own, but we also encourage parents and teachers to engage with children — ask questions to tease out their understanding of the world and provide guidance where and when it seems appropriate. We also encourage you to follow children's leads to encourage their interests in our magnificent world.

Table of Contents

2

4

6

8

10

12

14

16

18

20

 22

 32

 24

 34

 26

 36

 28

 38

 30

 40

Table of Contents

Table of Contents

42
44
46
48
50
52
54
56
58
60

Table of Contents

 62

 64

 66

 68

 70

 72

 74

 76

 78

 80

No not black and white and
oh so big.

But fluffy, cuddly, and cute;
a real charmer.

Standing, dancing, climbing;
a real whirligig.

But now a nap in the tree
seems in order.

*

A red panda is
a wonder in our world.

Sleeping through the winter - burr,

Kept warm with brown and fuzzy fur,

A mother watches each cub with care,

For any fool who might take that dare.

*

A family of grizzly bears is a wonder in our world.

The rain is raining all around,

It falls on field and tree,

It rains on the umbrellas here,

And on the ships at sea.

*

Rain is a wonder in our world.

Rain
by Robert Louis Stevenson, b. 1850

I am king of the daytime sky,
The lion of the air.

I soar and swirl and
swoop and dive.

No fish escapes from
my sharp eyes.

Up to my nest - up, up I rise,
To eat my dinner from on high.

*

A beautiful bald eagle is
a wonder in our world.

Flutter flutter flutter by,
Yet I am not a butterfly.

My coloring is not so bright,
I hide by day and fly by night.

Sometimes I fly into the light.
(Some people think I'm not too bright.)

*

A fuzzy red moth is
a wonder in our world.

Rumbling, grumbling, trembling.
Coughing, belching, smoking.

Hot molten rock streams out,
Forming land where once there was not.

What wakes that sleeping giant who has been quite dormant?

*

An erupting volcano is
a wonder in our world.

Carry you fast or carry you slow;

Carry you safely to and fro.

Rivers and valleys and oceans below;

Over and over and over I go.

*

A strong and beautiful bridge is
a wonder in our world.

Sunlit kisses - shady lanes,
I'm really happy when it rains.

Ancient forests - murky ponds,
Memories stored up in my fronds.

Graceful and feathery under trees,
Dancing on a summer breeze.

*

A field of ferns is
a wonder in our world.

In oceans deep and cold and wide,
I drift unthinking on the tide.

From here to there and back again,
With colors bright and squishy skin.

My tentacles know what to do,
I eat and float the whole day through.

*

A school of jellyfish is
a wonder in our world.

Twinkle, twinkle, little star,
How I wonder what you are!
Up above the world so high,
Like a diamond in the sky.

When the glorious sun is set,
When the grass with dew is wet,
Then you show your little light,
Twinkle, twinkle, all the night.

In the dark blue sky you keep
And often through my curtains peep;
For you never shut your eye
Till the sun is in the sky.

As your bright and tiny spark
Lights the traveller in the dark,
Though I know not what you are,
Twinkle, twinkle, little star.

*

A sky full of twinkling stars is
a wonder in our world.

The Little Star
by Ann & Jane Taylor,
sisters b. 1782 & 1783

A field of color that blooms in May;

Water it each and every day.

Watch them open in the sunlight;

Watch what happens when it's night.

*

Fields of red tulips are
a wonder in our world.

I love, oh, *how* I love to ride

On the fierce, foaming, bursting tide,

When every wave comes crashing ashore

Streaming back to where it came before.

*

Ocean waves are
a wonder in our world.

Storms bring me near.

Flashing lights from high above.

Clapping thunder is my escort.

It's said I don't strike twice,
but don't take that bet.

*

Flashing lightning is
a wonder in our world.

When I was down beside the sea
A wooden spade they gave to me
To dig the sandy shore.

My holes were empty like a cup.
In every hole the sea came up,
Till it could come no more.

*

A sandy beach is
a wonder in our world.

At the Sea-Side
by Robert Louis Stevenson, b. 1850

Sweet baby fishes come to me,
Swim safely in these roots of mine.

I sip salt water from the sea,
And make fresh water from the brine.

I stand between the sea and land,
My roots look like a Walking Man.

*

A stand of mangrove trees is
a wonder in our world.

Dark clouds forming,
Thunderstorming.

Dropping quickly from the sky,
Whirling twirling.

Spinning swirling,
Nothing's safe when I pass by.

*

A very scary tornado is
a wonder in our world.

Boats sail on the rivers,

And ships sail on the seas;

But clouds that sail across the sky

Are prettier far than these.

*

Clouds in the sky are
a wonder in our world.

The Clouds
by Christina G. Rossetti, b. 1830

Rarely in the light of day, darkness all around.

Drip, drip, drip for many, many years.

Looks like ice, but formed of rock.

*

A cave with stalactites is
a wonder in our world.

Twinkle twinkle from afar,
Yet I am not a twinkling star.

Speeding high above the ground,
Faster than the speed of sound.

You can watch me speeding by,
Like a comet in the sky.

*

The space station is
a wonder in our world.

The sky is painted gray;

And Earth is covered green;

But what is sandwiched in between?

A brush of color to paint the scene.

*

A double rainbow is
a wonder in our world.

All through bright days of June,

My leaves grow green and fair.

The soft and yellow silks soon,

Give way to ears for us to share.

*

A field of yummy corn is
a wonder in our world.

The sea! the sea! the open sea!

The blue, the fresh, the ever free!

Without a mark, without a bound,

It runneth the earth's wide
regions round;

It plays with the clouds;
it mocks the skies;

Or like a cradled creature lies.

*

Earth's ocean is
a wonder in our world.

The Sea
by Bryan Proctor, b. 1787, who wrote
under the name Barry Cornwell

Some little drops of water
Whose home was in the sea,
To go upon a journey
Once happened to agree.

A white cloud was their carriage,
Their horse a playful breeze;
And over town and country
They rode along at ease.

But, oh! there were so many,
At last the carriage broke,
And to the ground came tumbling
Those frightened little folk.

Among the grass and flowers
They then were forced to roam,
Until a brooklet found them
And carried them all home.

*

A babbling brook is
a wonder in our world.

The Raindrop's Ride
Anonymous

Moon, so round and yellow,
Looking from on high,
How I love to see you
Shining in the sky!

Oft and oft I wonder,
When I see you there,
How they get to light you,
Hanging in the air.

*

A full moon is
a wonder in our world.

Moon, So Round & Yellow
by Matthias Barr, b. 1831

A cute little home,

Is carried on her back.

Wherever she may roam,

She brings her little knapsack.

*

A hermit crab is
a wonder in our world.

At night when I go to bed;
I see the stars shine overhead.

And sometimes when all is right,
I see streaks of swirling,
glowing light.

*

The northern lights are
a wonder in our world.

Eggs make birds;
And birds make eggs;
Never to know which one came first.

Tiny ones will tweet!
Hatchlings soon to leap;
Emerged from sleep this happy new day!

*

A nest of baby birds is a wonder in our world.

O dandelion, yellow as gold,
What do you do all day?
I just wait here in the tall green grass
Till the children come to play.

O dandelion, yellow as gold,
What do you do all night?
I wait and wait till the cool dews fall
And my hair grows long and white.

And what do you do when your hair is white,
And the children come to play?
They take me up in their dimpled hands,
And blow my hair away.

*

Dandelion flowers are
a wonder in our world.

O' Dandelion
Anonymous

Hanging upside down with
hook-like toes;

Green gunk growing in the fur;

Living life at such slow paces;

A mother & baby with smiles upon
their faces.

*

A three-toed sloth is
a wonder in our world.

Water
Boiling,
Underground.

Blasts out,
With a
Crashing sound.

*

A hot, spewing geyser is
a wonder in our world.

Let's go for a sail

On the tail of a whale.

Gliding through oceans blue

Listening to songs we never knew.

*

A jumping whale is
a wonder in our world.

At evening when I go to bed

I see the stars shine overhead;

They are the little daisies white

That dot the meadow of the night.

*

A sky full of stars is
a wonder in our world.

Daisies
by Frank Dempster Sherman, b. 1860

How does the little busy bee
Improve each shining hour,
And gather honey all the day
From every opening flower!

How skillfully she builds her cell!
How neat she spreads the wax!
And labors hard to store it well
With the sweet food she makes.

*

Honey bees are
a wonder in our world.

How Does the Little Busy Bee
   by Isaac Watts, b. 1674

Once a home to some small creature;

With protection like that of armor;

Now when put up to my ear;

The sounds of the sea I still hear.

*

A seashell is
a wonder in our world.

Henry Ford could not even imagine,

Advances to his Model T invention,

Can travel without so much
as a driver,

Think they might get a bit bigger?

*

A self-driving car is
a wonder in our world.

A kingly lion rests in the sun,

Always ready to leap or to run.

With steely eyes and bushy mane,

He gazes out across the
golden plain.

*

A lion is
a wonder in our world.

Endless views show the
passage of time;

Layered rock reveal countless
ages past;

Cut by a river now far below;

Earth was changed. Will it
be again?

*

The Grand Canyon is
a wonder in our world.

Who has seen the wind?
Neither you nor I
But when the leaves hang trembling
The wind is passing by.

Who has seen the wind?
Neither you nor I
But when the trees bow down their heads
The wind is passing by.

*

A blowing wind is
a wonder in our world.

The Wind
by Christina G. Rossetti, b. 1830

Starting small but growing wide,

Moving 'cross the ocean's tide.

Howling winds and driving rain,

A one-eyed monster hurricane.

*

A hurricane is
a wonder in our world.

Heart free, hand free,
Blue above, brown under,
All the world to me
Is a place of wonder.

Sunshine, moonshine,
Stars, and winds a-blowing,
All into this heart of mine
Flowing, flowing, flowing!

*

Earth is a wonder.

The World of Wonder
by William Stanley Braithwaite, b. 1878

Adventures from nVizn Ideas

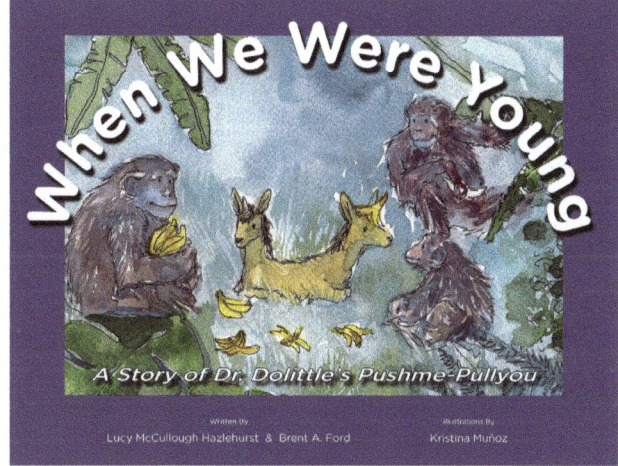

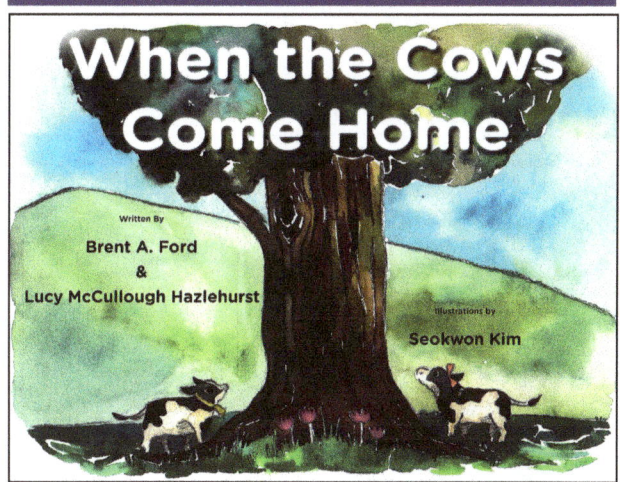
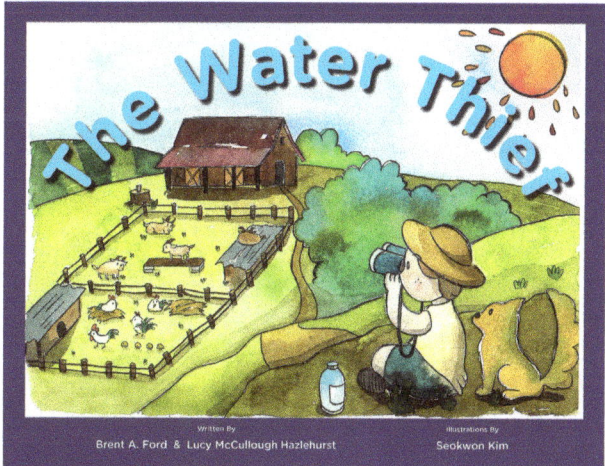

Ozzie & Alina Adventures

Updated Classics

Science & Nature eBooks

Interactive eBook

Variety in the Animal World

www.ingramcontent.com/pod-product-compliance
Lightning Source LLC
Chambersburg PA
CBHW040223040426
42333CB00051B/3419